Biology

Science of Life, Cell Theory, Evolution, Genetics, Homeostasis and Energy

2nd Edition

By Robert Meeks

Robert Meeks

© Copyright 2016 by Robert Meeks – All rights reserved.

Robert Meeks

Table of Contents

Robert Meeks

Introduction

First of all, I would like to thank you for choosing this book, "Biology: Science of Life, Cell Theory, Evolution, Genetics, Homeostasis and Energy."

Do you get confused by all the talk about genetics, evolution, cell theory, homeostasis, DNA and so on? Well, you needn't worry about feeling the same ever again. This book is a simple guide that will answer some of the most basic questions about the above-mentioned concepts, and towards the end of this book, you will be able to get a better understanding of life as a whole on this earth and the process of evolution.

You will learn more about the various external factors that brought about Biology as a study and how we can see and use it in your daily life. Knowing the foundation of how living creatures are made and how they evolve will leave you feeling that every living creature is a wonder to behold.

We will also discuss the human body and its functions. There is nothing more fascinating than to discover the inner workings of your own body. Knowledge is power and knowing how your body works and what it takes to thrive will give you the ability to give it more of what it needs. Furthermore, we will give you an idea how biology relates to other fields of study such as psychology, chemistry, and others. We will also delve into the background of biology as a study and how it relates to the modern individual.

For this reason, I hope this book will serve as a useful and educational resource for you. I am sure that it will enrich your knowledge of biology, your body and the science of life.

Without any further ado, let us get started! Happy reading!

Chapter 1 – Basics of Biology

To put it simply, biology is known as the science of life. The structure, function, growth, as well as the distribution and the evolution of living organisms are the areas that are studied by biologists. There are nine different fields that are considered to be the umbrella fields of biology.

The first one is biochemistry; this is the study of all the material substances that the living beings are made up of. The second one is botany; this is the study of all plants and even includes agriculture. The third area is cellular biology, and like its name suggests, this is the study of the most basic of units that make up the living beings, the cells. The fourth field is ecology, which is the study of the interaction that takes place between the organisms and the environment. The fifth one is the study of the origins of species, as well as the changes that have taken place in life over a period of time, and this is referred to as evolutionary biology. The sixth area is genetics, and this is the study of heredity. The seventh concept is the study of various biological molecules that exist, and it's referred to as molecular biology. Physiology is one of the eight areas, and it is the study of the various functions performed by organisms and their different parts. Finally, the ninth concept that comes under the umbrella of biology is zoology, which is the study of different animals and their behaviors.

Biology, in fact, is a vast subject and what makes this a little more complex is that all the above-mentioned fields are overlapping. For instance, you cannot really study much about zoology if you don't understand the concepts of evolution, physiology, genetics and ecology. Similarly, each of these fields is intricately related to the other fields.

Additional Information

The word Biology comes from the Greek words of bios meaning life and the suffix logy meaning "science of." This suffix is rooted from the Greek word leigen, which means to select or to gather. In essence, biology is in totality, "to study of life". It answers the questions "where does life come from?" Thomas Bedoes introduced the term itself, in the year 1799 followed by a group of fellow scientists who lived in the early 1800s as well.

What is interesting is that before recorded history, there were many terms used to describe animals, plants and other organisms physically. There were many different unorthodox ways on how to study them. Another branch of biology called "natural history" touched on these various terms. However, the study of natural history includes non-living organisms such as minerals and various chemicals.

In the middle ages, this unifying framework related to the natural history was known as the Scala Naturae. This concept is known as the great chain of being. While biology touched on the physical aspect of living, this particular branch of biology touched on the spiritual or metaphysical side of biological studies. This is where the reason behind an organism's existence is further explored. It answers the question why does an organism exist. In short, while biology deals with the physicality of an organism's existence, natural sciences, and history tend to deal with the spirituality of the concept. That being said one cannot exist without the other. This is why it is important for us to learn biology and the various concepts related to the study as a whole.

Basic Framework

There are various known branches of biology. Branches are correlated to one another. You will see how it relates to other scientific concepts as you continue reading.

You can only do this effectively if you understand the basic history behind biology in relation to the physical world. If you explore it further, you will also learn about the five basic principles that govern the subject of biology. This section will discuss in passing these basic principles as they operate in the extended biosphere. You will also learn how they apply to everyday living as a whole.

All the different branches of biology can be organized under five basic principles that govern the understanding of all living beings. The five basic areas are cell theory, heredity, energy, homeostasis and evolution. In this book, we shall take a look at these five essential concepts in the coming chapters. But for now, let us take a brief look at these concepts.

Cell theory is divided into three parts for getting a better understanding of it and these are:

1. Cells are the most basic units of life

2. All the living creatures that exist are made up of cells

3. Cells arise from the pre-existing cells.

So, to put it simply, cells are the building blocks of life.

Heredity is the study of DNA and various genetic codes that form the basic structure and the functioning of cells.

Energy is the basic requirement of life. Energy flows between different organisms and the organisms and the environment around.

The fourth concept is that of homeostasis and this means that all the living beings need to be in a state of balance, that is they must be in equilibrium and this equilibrium must be between the organisms and the environment as well.

The fifth and the final concept is that of evolution and this is the concept that brings together all the other areas of biology, this is the change that takes place over a period of time and is responsible for biological diversity that exists.

History Of Biology

Our fascination with biology isn't something new; in fact, evidence has been found that our ancestors also studied about the various animals they hunted and knew the locations of different species of plants that they gathered for food. The advent of agriculture is perhaps the first major leap for humanity. Medicine has been a crucial part of different civilizations and the earliest known texts related to medicine have been traced back to China to 2500 B.C., Egypt in 1800 B.C. and Mesopotamia in 2112 B.C.

The recorded history of biology also traces back to ancient civilizations of the known world. Biology as a concept didn't exist until the late 19th century. As mentioned earlier in the text, sciences related to biology emerge from medical traditions and natural healing. The medical practices related to the concept even dates back to early Egyptian medicine, Ayurveda, and other Greek and Roman practices that were well known at that time.

During the classical times, Aristotle has been considered to be the first person that started the practice of scientific zoology. Aristotle is believed to have performed several extensive studies regarding marine life as well as plants. Theophrastus was his student and he was the author of one of the earliest known texts in the West regarding the structure, life cycle and the various uses of plants. Galen was a Roman physician and he made use of the expertise he gained while patching up gladiators for the arena to write about the different surgical procedures in 158 A.D.

During the period of Renaissance, Leonardo da Vinci, one of the most famous names till date, had risked censure by actively taking part in human dissection and then making extremely detailed anatomical drawings of the human body. These drawings are still considered to be some of the most beautiful and perfect representations of what the human body actually looks like. With the invention of the printing press, it became really easy to record as well as disseminates information. Leonhard Fuchs was a German botanists and one of the first books on biology with illustrations was written by him in the year 1542. Binomial classification that is made use of even today was started by Carolus Linnaeus in the year 1735.This practice is the usage of Latin names for grouping various species based on their characteristics.

In Europe, because of the Renaissance, there was a well-known revolution regarding biological studies and thought. This is because of the revival of empiricism as part of the medical studies. Early discoveries of new organisms also occurred during this time. The study of biology branched out into the observations of Physiology because of the further exploration of these newly discovered organisms.

The invention of microscopes in the year 1665 opened up a completely new dimension for scientists. In the year of 1665, Robert Hooke had made use of a simple compound microscope, the most rudimentary microscope, for examining a thin sliver of cork. This is when he noticed that the plant tissue is actually made up of tiny rooms and partitions, these so called rooms like structures reminded him of the rooms that the monks used to live in. And he came up with the name "cells" for referring to these units. In the year 1676, Anton von Leeuwenhoek had published the first drawing of the living cells that an organism was made up of. Later on in the year 1839, Theodore Schwann had added a very important piece of information by proving that even the animal tissue is made up of cells. It was only during the Victorian era and the whole of 19th century that natural science had become all the rage. There were thousands of species that were discovered and described in extreme detail by adventurers and backyard botanists as well as entomologists. In the year 1812, it was Georges Cuvier who had described fossils and had come up with the hypothecation that the Earth we are living on had gone through different phases of creation and destruction over due course of time. Charles Darwin had in the year 1859 on the 24th of November, published a text titled "On the Origin of

Species" that changed the perception of life. This text proclaimed that all the species are interrelated and that we had all come from the ancestral forms depending upon our ability to adapt to the environment.

The attention of the world was directed mostly towards questions related to macroscopic organisms; there was a quiet monk who had been silently investigating the manner in which traits are passed on from one generation to the other. This monk was none other than Gregor Mendelis, who is considered to be the father of genetics and the basis of genetics have been developed on the texts that he had published in the year 1866. Though they had gone unnoticed at the time of their publication, it was only in the year 1990 and the ones to follow that his texts had gained prominence.

The 20th as well as the 21st centuries can be referred to the future generations as the era when the biological revolution had started. Watson and Crick had explained about the concepts regarding the structure as well as the functioning of DNA in the year 1953 and this had a far reaching effect on various fields of biology and has managed to touch different aspects of our lives directly as well as indirectly. The field of medicine has also undergone rapid transformation and there are different therapies for treating patients that have been specifically developed keeping in mind the genetic blueprint of the patient by combining biology and technology as one. There are endless possibilities of all the advancements that the human race would make.

Summary

To summarize, some prominent biological and scientific figures during this time are as follows:

- Andreas Vesalius
- William Harvey

- Carl Linnaeus

- Leonardo Da Vinci

- Aristotle

These world renowned scientists contributed to the further development and study of the various different branches of biology such as Anatomy, Physiology and Botany. Aside from giving contributions to the said scientific branches, these people and their cohorts were pioneers when it came to many different developments in the study of various organisms. It was also during the time of the Renaissance that classifications regarding biological organisms as well as the recording of discovered fossils came in to fruition. These experts also contributed to the development of industrial equipment that further enhanced the study of biology.

Now that you know more about the beginnings and history of biology, we will now focus on the basic tenets of biological study. In the next chapter, you will find out more about cellular theory. You will also learn how it affects the internal structure of the physical body as a whole.

In summary, the history of biology as a concept and a study began from the early 16th century up to the present. This means that there is a wide variety and a vast amount of information that you will be able to discover. With all the prominent figures and interesting scientific concepts encompassing the subject, you will certainly figure out the importance of biology and how you can apply it to your daily living in the end.

Chapter 2 – Cell Theory

What is Cell Theory?

As we all know, human beings have millions of tiny cells in their bodies. No human can exist without them and are the most important parts of our bodies. These cells are microscopic in nature and you can only see them if you use a microscope. Our bodies are made of trillions of cells and scientists claim that we might have around 100 trillion cells in our bodies.

There are four basic concepts that you must understand about cells and they are as follows.

- All living beings have cells

This is the first and most important theory that you have to know about cells. Each and every living organism will have cells regardless of whether they are small or big. So right from an elephant to an ant, every living being has cells in their bodies. If

Cells: All living things are made up of cellular matter. It is a building block of earthly existence. It is the minutest of organization units in a living organism. This is where you can find the DNA strand of humans as well as animals. Therefore, the basic cellular makeup can help determine heredity as well as the physical traits of certain organisms. Cells can multiply through the process of mitosis or cell division to form the various tissues and muscles in the body.

Furthermore, our cells are responsible for many of the bodily functions that we experience on a daily basis. From breathing to the internal functions of our organs, we need cells to survive. So let us not underestimate the power of the human cell. We won't be able to exist without it.

Because they make up living things, cells can also be considered as the living units inside the body. On this account, they also undergo the process of ingestion and cellular aspiration to survive. This is part of the process of cell division, which helps perpetuate the existence of cells inside the body. This also proves the third point of the cell theory.

- Cells are the smallest living organisms and can perform all functions of life

These cells are tiny yet perform all possible functions of life. So don't go by their size, as they can do everything that you can do. These cells are pretty independent and will perform their duties all by themselves.

- Cells come from pre-existing ones

All cells reproduce to create new cells. Thus, the new cells in your body will have to come from existing ones. This happens through a process known as cell division, which causes these new cells to be produced. These cells contain genetic information, which they pass on to the next batch. So, your cells are tiny information storehouses that pass it on to subsequent batches and so on. Sometimes, little details can get left out.

- The nucleus is the heart of the cell

Just like how we have a heart, which commands our cell's functions, cells too have a heart known as the nucleus.

The Nucleus Defined

It is the central part of the cell that contains the human genetic material or deoxyribonucleic acid or DNA. It is a small organelle enclosed in a membrane that can be found in eukaryotic cells (cells protected by membranes). Usually, eukaryotic cells only have one nucleus. However, in certain organisms there can be multiple occurrences of nuclei in one cell. These will be further discussed as the chapter continues.

These form the 4 main theories of cells that you must know to understand them better.

Apart from these, there are many other theories that are secondary in nature. For example, these cells consume chemicals to remain alive. It is extremely important for cells to get what they want, especially if they are required to divide.

As research continued throughout the years, additional tenets have been widely accepted and added to the ones above. These form the Modern Cell Theory.

- Cells contain hereditary information, which is passed from cell to cell during cell division.

- All cells are basically the same in chemical composition.

- All energy flow (metabolism & biochemistry) of life occurs within cells.

History Of Cells

Cells were first discovered in early 1600 by Robert Hooke, who observed leaves under a microscope and found tiny little cells that performed many functions.

He continued his experiments for a long time and also described them in great detail to aid with future references.

Later, two scientists, Nehemiah Grew and Marcello Malpighi studied animal cells to understand their structure better. They were keen to know what these cells looked like and the real purpose that they served. They found that these cells looked much different from leaf cells.

A few years later, protozoa and other microorganisms were discovered, which further opened up the study of cells. Cell theory began to take shape, and suddenly, there was an outpour of studies because everybody was keen on understanding what these cells really did within these organisms. It was clear that they had a significant purpose in the human body, but their true value was yet to be established.

As time passed, people understood that these cells are life giving and are pretty much responsible for all bodily functions to be carried out with ease.

These cells are what allow us humans, plants and animals to survive.

Cells And Energy

Cells are extremely important and you can only survive if your cells are multiplying correctly. If they are not, then illnesses are sure to come about.

All cells have a free flowing energy inside them, which causes them to function optimally. The energy inside them is pretty much a constant and is fueled by chemicals.

These chemicals are what nourish and enrich our body's cells.

The Correlation

Cells have significant amounts of energy that is necessary for them to survive. Without cellular energy, these organisms will not be able to function normally. As mentioned earlier, the energy comes from a combination of natural chemicals produced by the body through the various nutrients that are taken in every day. It is through this that the body is able to produce the necessary oxygen to allow the organs to work properly without fail. Without the proper nutrients, they will not be able to produce oxygen and in turn, render the body or some organs unable to function.

Types Of Cells

There are two main types of cells namely the Eukaryotic and prokaryotic cells.

The eukaryotic cells have a defined nucleus. The nucleus is the heart of the cell and what controls most of its activities. The nucleus is contained within a membrane that is separated from the rest of the cell's contents. Such cells belong to animals, plants etc. The nucleus contains the distinct DNA.

On the other hand, the prokaryotic cells do not have a distinct nucleus. The DNA is not distinctly defined and everything merges in together. Bacteria are said to be prokaryotic cells.

These cells can exist in our bodies, mainly in our digestive tracts. The human body is said to contain more bacteria than regular eukaryotic cells.

Another important point of distinction between the two is that eukaryotic cells contain many organelles such as ribosomes while prokaryotic contain just a few.

Cell Reproduction

Cells reproduce in three main ways namely binary fusion, mitosis and meiosis. Binary fusion refers to a single cell producing two other identical cells. It is also known as cloning. Mitosis is the process of parent cells splitting into distinctive daughter cells. The parent cell will always pass on some of its characteristics to the two new cells. Meiosis refers to a sexual form of cell division. Two cells sexually produce a third cell through a process known as meiosis.

The Process of Mitosis

In this section, we will discuss in detail how cell division occurs.

What exactly happens to a cell when it multiplies and how can you externally help speed it up if necessary.

Mitosis refers to the process of cell division that allows the regeneration of new cells, which can therefore foster the re-growth, and healing of certain body parts. Cell division happens when the nucleus of one cell breaks off into multiple identical nuclei called daughters. Mitosis can be divided into four official phases. These are as follows:

- Prophase

- Metaphase

- Anaphase

- Telophase

After the four initial phases of mitosis the next process is called cytokinesis. This is the event where two fully developed cells completely divide and separate from one another. This completes cell reproduction.

Meiosis

After the cells naturally divide as part of mitosis, there is another kind of division that occurs in the chromosomes. This is known as meiosis. It can be defined as a form of cell division that produces haploid sex cells (gametes) directly from a root diploid cell. This process replicates one strand of DNA subsequently followed by double nuclear and cellular divisions.

Haploid and Diploid Gametes

A haploid sex cell refers to a gamete that contains a single copy of chromosomes in one DNA. On the other hand, a diploid sex cell contains two copies of each chromosome in one DNA.

In short, mitosis is a process occurs in cellular matter while meiosis happens in chromosomes.

Parts of the Cell

The two main parts of the cell include the cell nucleus and the cytoplasm. These two parts are covered by a thin cell wall called the membrane. This serves as a protection against the external environment of the cell. Below are some of the additional parts of the cell that the need to keep in mind in order to understand the human biology to the fullest.

- **Ribosomes:** This is where the RNA or ribonucleic acid is directly converted into protein. This process is referred to as the protein synthesis. This is a significant process in the life cycle of the cell because it helps produce an equally significant strand of heredity that will determine genetics in the future. In addition, ribosomes can be found floating around and inside the cytoplasm.

- The **Endoplasmic Reticulum** referred to as the transport system of the body that will allow certain molecules to go in specific directions.

- **Lysosomes:** This serves as the digestive system of the cell. It breaks down nutrient molecules in the body into the basic components. Because of this, it renders the nutrients easily digestible for the cell.

- **Microtubules** serve as a temporary filling for organelles during cell division. It protects the smaller parts of the cellular matter from being damaged while this cell is still dividing.

- **Vacuoles** are organelles with a single membrane that is located in the external portion of the cell.

There are many other smaller parts of cells that are worth mentioning. These are as follows:

- Golgi bodies

- Chloroplasts

- Mitochondria (sing. Mitochondrion)

Note that chloroplasts can directly be found in plants alone. This is where the DNA of plants is located. The golgi bodies alter molecules to create microscopic sacs inside the cells called vesicles. The mitochondrion on the other hand, is the main energy source of the cell. It is composed of both an inner and outer membrane.

The main purpose of cell division is to transfer genetic matter from one cell to the other. This is how living organisms reproduce. It is definitely something that is worth exploring even further through additional study and research.

Cell Life

Cell life varies and depends on where the cells live. Some might live for a few days while some others will live for a year. Certain immune system cells live for a few weeks while the ones present in the pancreas can live for a year.

Cells have the tendency to self-destruct if an illnesses or disease affects it. This process is better known as apoptosis. It is a natural process that regular cells undergo to keep the body healthy.

If a cell is unable to undergo apoptosis then it will result in cancer.

The Lifespan of the Cell

As mentioned above, the life span of a cell can range from a few days to many years depending on how healthy the cell is. The human skin is a great example of the healthiest group of cells that the human body has. Why do I say that it is healthy? Because the skin can constantly regenerate without difficulty.

There are certain internal environments that can be detrimental to the life expectancy of a human cell. As I said, it all depends on how strong the immune system is. If you have an illness, this slows down cellular metabolism, which can directly affect regeneration and the health of the host. This is why you have to boost your immune system as much as possible. It will certainly help maintain the regeneration process and prevent bacteria from attacking your immune system constantly.

How to Boost the Immune System

You can do this by stocking up on your vitamins for the body. Vitamin supplements are good, but I would still recommend getting it from the natural sources. Organic fruits and vegetables are the key.

You also have to remember that some of the organs of the body are not able to regenerate new cells. This means that when the organ is damaged, chances are you won't be able to replace it internally. An example of this non-regenerative organ would be the liver.

There are also certain illnesses that attack the cells of the body as well as the immune system. The Human Immunodeficiency Virus or HIV is one such culprit. If you have the HIV in your system, your cells will not be able to regenerate antibodies that can combat the virus. This is mainly because the virus attacks the immune system first and when the immune system goes down so does the defense of the body against other viruses and bacteria that may harm the good cells in the end.

This is why it is important for you to take care of your immune system. As they say, prevention is always a thousand times better than finding a cure. If you are able to do this successfully, you would then prevent further difficulties along the way. Therefore, if I were you, I will definitely find a way to boost my immune system as soon as possible. If you are having difficulty finding ways, here are some additional specific tips that you should follow:

Add nuts to your diet. This will not only increase your memory, but also boost up your immune system significantly. Studies also show that drinking lots of water will be able to increase the nutrients in your body and replenish those that you have already lost. Daily exercise is also necessary for you to build up your tolerance against physical ailments in the future.

Knowing these tips can definitely give you what you want in terms of protection against possible cellular degeneration in the long run. Having discipline is only one of the best ways for you to fight off certain diseases. You will definitely have a longer life if you're disciplined enough to take care of your body.

Studying Cells

If you wish to remove some cells from your body to place under a microscope then you can use a Q-tip to rub on the insides of your mouth and you will be left with a lot of skin cells. These are generally dumped by the body at the end of the day and new ones are formed the next day.

How Are They Studied?

Attempting to look into the cellular makeup of the body has come a long way since the days of Robert Hooke. He was the British scientist who started describing physically cells in the mid-1600s. Today, many technological advances have allowed scientists to see into of the innermost parts of the cell in real time. One of these technical pieces of equipment and advances is **the microscope**.

The Microscope

One such equipment that played an important role in further studies is the microscope. The development of the microscope during the renaissance and industrial age paved the way for the discovery of the basic unit of human existence known as cells.

According to some historians, the Dutch spectacle maker Zacharias Janssen invented the compound microscope in the year 1590. This particular kind of microscope utilizes lenses and light to increase the size of an image. It is also known as the optical or light microscope.

There is one other kind of microscope that has been invented. This is the electron microscope. This specific type of microscope uses electric beams to create an image of the object being looked into. It is extremely capable of magnifying the specimen even more and has much higher resolution. Because of this, detailed pictures of an extremely minute landscape are now possible.

The earliest form on the microscope would be the magnifying glass. It allowed a specimen to be seen about 10 times larger. This is why you should never underestimate the power of the magnifying glass. Without the magnifying glass, there would be no precedent to the current technological advancement that we have today.

With the help of the microscope, they will be able to view the internal cellular makeup of the body, which then will allow them to find out what may be wrong with the physical form of a human being. They can also use these pieces of information to figure out how cells may react to environmental stressors both inside and outside the body. Some of these environmental stressors include rising temperatures inside the body or toxins introduced into the bloodstream. After gathering all the data, scientists and doctors will be able to use other sophisticated tools to further explore and understand the human body. This can be done through further biochemical testing

Cell Death: Apoptosis and Necrosis

When there is life, there always must be death. The same saying can be applied to the cellular environment. This cell death can also be called apoptosis. This is the pre-programmed instruction of the cell to self-destruct in order to protect the body from further damage and harm in case it is attacked by foreign entities that seek to lower the immune system such as a virus or bacteria.

Aside from being a protective mechanism, apoptosis is a process that helps keep the body in shape in terms of maintaining the form and function of the external and internal organs of the said physical form. It also helps fight off diseases by destroying infected cells in the body and preventing the spread of infection through the bloodstream.

There is another kind of cell death known as necrosis. In this particular phase, everything happens all of a sudden. Necrosis can happen if the physical body experiences sudden injury or an exposure to toxic chemicals that may mutate the cells and cause irreparable damage to the organs.

Speaking of irreparable damage, the moment that the organs become inflamed or infected, diseases will come about and eventually cause incurable diseases such as cancer or Acquired Immuno Deficiency Syndrome.

Understanding Human Health and Existence through the Study of Cells

How does the study of cells contribute to the human understanding of overall health and disease? Studying cells help in such a way that you would be able to discover how they work and therefore be able to protect them accordingly by finding out how you can prevent infection from attacking healthy cells inside the body.

Cellular studies will help you discover what you can do to keep your own cells healthy. By understanding the basics of cellular studies, you begin to be able to understand what many people strive for their entire life. As the animals, mammals, and humans that we actually are, we all have a natural desire to survive. By understanding biology, you are really beginning to understand how to increase your survival by years. This has been shown recently, with a large increase in lifespan over the last few decades. Much of this increase in lifespan is due to society as a whole understanding the very basics of what you are learning here, and then applying it to their health.

Not long ago, simple diseases could have been life threatening, but with the advances in understanding, came huge gains in healthcare. For centuries humans did not understand the basics of biology, and by studying natural history, we can see that our lifespan was rather short. Years ago, without the vast knowledge we now have, a simple cut on the foot which became infected could be life threatening. Without the microscopes, medication, and advanced technology that are available currently, the only way humans survived was really through trail and error. Meaning to learn how to treat a simple foot infection, for example, we had to use trial and error, with that person's foot and ultimately with their life. We were essentially learning basics of health, by seeing whether someone survived, when a particular remedy was used.

Physicians and researchers are now infinitely more aware of how certain diseases work, even down to the cellular level. Due to that increased understanding we are now able to create medications and therapies that are specific to a particular disease. Many medications now are actually even specific to a certain type of cell, or even a certain part of a specific type of cell. It is absolutely amazing how understanding basic cellular biology has directly affected human health and existence during our lifetime.

Personally, you may be trying to understand how you can use biology to affect your own health. By understanding it, you'll be able to answer questions like: What food to eat? What are the advantages and disadvantages of drinking lots of water for the cells? What kind of lifestyle should you adopt in order to take care of your body and cellular makeup without difficulty? These are just some of the major questions that can be answered by further cellular studies. So if you are looking to improve your health, definitely continue to research and learn more about how your body works, and how certain cells (which make up certain foods, etc.) affect your physical self. I am sure that you will not regret doing so.

Chapter 3 – Evolution of Life

Introduction To Evolution

Before getting started with the different concepts of evolution it is important to understand the meaning of evolution. Evolution is the decent that is combined with modification. This definition is the amalgamation of small scale evolution like the changes in the frequency of genes in the population of one generation to the other and the concepts of large scale evolution like the descent of a different species from a common ancestor tracing back to generations. Evolution is the concept that will help you understand the history of life, as we know it.

Biological evolution is not a simple change that took place over time. There are various things that change with the passage of time. Trees shed their leaves, mountain ranges erode, but these instances aren't the examples of biological evolution because they do not involve the descent of genetic inheritance. The main concept of biological evolution is that all the life that exists on Earth can be traced back to one common ancestor. A simple example would be that you and your cousins have descended from a common grandparent. When the process of descent is combined with modification, the common ancestor of ours has given rise to all the diversity that has been documented in the form of fossils and also everything around us today. Evolution states that we are all related to one another. The implications of this statement are numerous.

Life on earth has existed for millions of years. Tiny bacteria are said to have been the first living organisms on Earth and are regarded to be our ancestors. Many argue that this is not a real possibility, as humans could not have descended from tiny microorganisms. There should have been something bigger that evolved into human shape. However, according to experts in the field of evolution, it is a must that everything on earth came from the same source and gradually evolved into different beings. Despite several years of studies, nobody knows for sure as to where these tiny bacteria, with rod like projections emerging from their bodies, might have come from. Some assume that they are not from earth at all while others believe they self-generated owing to Earth's climate being suitable.

Research is still being conducted on these "bacteria fossils," and scientists are trying to establish a co-relation between these prokaryotic cells and the beginning of the universe. Better known as cyanobacteria, they were bluish green algae that formed on rocks. These bacteria contained prokaryotic cells, which were much simpler in nature than eukaryotic ones.

Earth is said to have been in an inhabitable state until carbon and water were created. Both of these are extremely important for all organisms to survive.

Scientists state that the first real breakthrough in the world of evolution was the creation of lipid bubbles. These bubbles were double walled and would reproduce by themselves. This is comparable to cell theory and how existing cells reproduce to form new ones. Soon, stromatolites came into being, which are best described as "large structures formed by many cyanobacteria combining together". Between 2500 m.y. and 544 m.y., animals with eukaryotic cells came into existence. Their cell structure was a bit more complex as the nucleus was well defined and there were many other small yet distinct components present within it.

Subsequently, soft-bodied animals came into being. These animals are said to have been different from sponges. They contained a skeletal framework. Between 544 m.y. and 505 m.y., the Cambrian explosion took place, which caused millions of small animals to take shape and inhabit Earth. These animals mostly consisted of trilobites, brachiopods, ostracods and sponges. Each one had a distinctive feature to them and was pretty simple to tell them apart. Between 505 m.y. and 440 m.y., there was an emergence of fish. And once they arrived, plants soon followed suit. Most of these fish had vertebrates, which meant that they contained distinct backbones.

Between 410 m.y. and 360 m.y., several types of insects came into being. These insects had a tough task at hand, as they had to come up with ways through which they could control water loss and increase oxygen intake, owing to a transition from water to land. This era was better known as the Devonian period. Subsequently, during the carboniferous period, there was the rise of the reptiles. These reptiles began to lay eggs and reproduce. This era was marked by the rise of amphibians that could survive both in water and on land. During 286 m.y. and 248 m.y., there was a marked expansion of the reptile kingdom that existed side by side with a plethora of insects and other small animals.

Theory Of Evolution

Like mentioned earlier, Darwin had published the book titled "On the Origin of Species" in the year 1859 and this is when the theory of evolution by natural selection had been discussed for the first time ever. This is the process by which all the organisms have undergone changes over a period of time due to the changes in physical as well as behavioral traits that can be inherited. There are some changes that help an organism to get better adapted to its environment and also help in their survival and reproduction.

This theory of evolution by natural selection is considered to be one of the most substantial theories of science that is backed by a variety of scientific disciplines like paleontology, geology, genetics and even developmental biology. This theory is based on two main parts. The first part suggests that all life on Earth is connected and related to each other and the second part is that the diversity in life is due to the modifications that are a result of natural selection wherein some traits and environments have been preferred over the others. This theory, to state it simply is all about descent combined with modification and this theory is also referred to as survival of the fittest. In this context, the word fitness doesn't refer to the strength or the athletic ability of an organism, but it refers to the ability of the organism to survive.

In the first edition of the book that was published by Charles Darwin in the year 1859, he had speculated over the concept of how natural selection would make a land mammal turn into a whale over the course of time. To explain this concept, he had to make use of a hypothecation that the North American black bears have been known to catch insects in the water by swimming with their mouths open.

Darwin had speculated that perhaps due to natural selection, the bears started adapting to their aquatic structure and habits till the point where they had been evolved into a whale. This speculation of his wasn't received well by the public and the vehement ridicule that he was subjected to made him decide to do away with this speculation in the following editions of the book. Though the scientists today would agree that Darwin indeed had the right idea but he chose the wrong animal. Instead of a bear he should have taken a look at cows and hippopotamuses. The story of the evolution and origin of whales is perhaps one of the most fascinating tales of evolution and the best examples of natural selection.

You must be what natural selection is. Natural selection can change a species in minute ways and this causes the population to change its color or size over several generations and this process is referred to as microevolution. There is so much more that natural selection is capable of. When given sufficient time and changes, natural selection is capable of creating a new species altogether and this is referred to as macroevolution. This can turn dinosaurs to birds, and amphibious mammals to whales like we have discussed earlier and the best example of all would be the evolution of human beings from apes. Let us take whales as an example; making use of evolution and the process of natural selection, biologists can now explain the manner in which whales might have transitioned from land to water. For instance, the evolution of their blowhole could have happened in the following manner.

Random genetic changes would have resulted in the placement of the nostrils of a whale father back on its head and all those whales that had inherited this change would have been more suited for the marine life since they didn't have to surface completely in order to breathe. Such animals would have had been more successful and the same would have been passed onto their offspring as well. With each passing generation, the nose would have been moved farther back. The other body parts of the early whales would have also gone through a lot of transition. The front limbs would have been replaced by flippers and the hind limbs would have disappeared altogether due to their reduced usage. Their bodies would have gotten more streamlined and then they also would have developed tail flukes that will help in their movement through water. Not just this, Darwin also described that natural selection also occurred depending upon the success rate of the organism for attracting a mate and this process is referred to as sexual selection.

The two examples of this are the bright and colorful plumage of a peacock and the antlers of a male deer. Darwin wasn't the first or the only scientist who had come up with a theory of evolution. Jean Baptiste Lamarck was a French biologist and he had come up with the idea that an organism is capable of passing on its traits to its offspring, but there were some details that he had gotten wrong. Around that time Alfred Russel Wallace, a British biologist also came up with a theory of evolution that was based on natural selection.

When we think of this from a modern perspective, Darwin didn't have any idea about genetics. He just observed the process of evolution without knowing anything about the mechanism behind it. All this had come later on, when it was discovered the manner in which genes are encoded in our different biological as well as behavioral traits and the manner in which these genes are passed on from parents to their offspring. Modern evolutionary synthesis is the phrase that is coined for understanding the incorporation of genetics in Darwin's theory of evolution.

All the various physical and behavioral changes that assist in natural selection take place at the level of DNA and these changes are referred to as mutations. Mutations can simply be understood as the building blocks for evolution. There are different causes for mutation and these could be due to any random changes in the DNA repair or replication or even due to chemical or radiation damage. Most of the times mutations can either be harmful or neutral but only in rare circumstances this process can be advantageous for the organism. Only when this happens, such a mutation will be more prevalent in the future generations and will spread through the population slowly. And in this manner, natural selection acts as a guide for the evolutionary process and helps in preserving as well as adding the beneficial mutations to the species while rejecting the negative ones. Mutations indeed are very random, but selection for these mutations isn't random. But this isn't the only mechanism through which organisms can evolve. For instance, the genes can also be transferred from one population to another one when the organisms tend to migrate or immigrate and this process is referred to as gene flow. Genetic drift is the frequency at which a particular gene can change at random.

Even though the scientists have predicted the appearance of the early whales, the absence of required fossil evidence has been taken as a proof by the creationists that the process of evolution hadn't taken place at all. Creationists have always mocked the idea of a whale that could walk on land, and scientists have been trying to find evidence to back their claim since the early 1990s.

During the year 1994, an important piece of evidence presented itself in the form of the fossil remains of Ambulocetus natans; its name is a literal translation for walking whale. The forelimbs of this creature had fingers as well as small hooves whereas its hind feet were just huge given the size of this creature. This fossil gave the evidence that this creature was adapted for swimming but it could also manage some clumsy movement while on ground, almost like a seal. When this creature swam, it perhaps moved in a manner similar to that exhibited by an otter and made use of its hind legs and tail for swimming. Whereas modern whales propel themselves through water by using their tail flukes, unlike the ancient creature that had a whip like tail and hind legs for propelling itself forwards with the required force. In the recent years there have been accounts where the number of so called missing links in the chain of evolution has been discovered.

In spite of all the evidence that the fossils have proved, as well as genetics and other fields of science, there are still some people who question the validity of the theory of evolution. There is a lot of controversy that surrounds this theory. But the main stream scientists don't really see any controversy associated with this because the evidence that they have been finding points in the direction of Darwin's theory.

Now that you know all about human evolution, we will discuss some of the terms that you might find confusing. Some of these terms are as follows:

Homo – this term is rooted from Latin, which means human. The word itself comes from Indo-European roots of lexis that generally pertain to earth. Primarily Carolinus Linnaeus chose it. He and other famed scientific personalities of his time believed that humans are directly linked to apes due to the morphology and physical similarities of the said animal to the human physical component. This is basically the precedent of Darwin's theory of evolution that has been known in the scientific world for many years already.

Human – this word comes from the Latin word humanus, which is considered as the adjective form of homo.

Fossils and Genetics

Because of further archaeological finds in the subsequent years, fossils were discovered to support the theory of evolution. On this account, evolution was further explored and linked to other fields of science such as chemistry and genetics. Because of the minimal differences between the Neanderthal fossils and bones to the current human skeletal framework, genetics and the evolution of all the cellular makeup were deemed as the cause of the stark similarities and differences.

The Psychological Effects of Biology on the Personality

In terms of evolution, it is important for you to learn all that you can about genetics in connection with biology. You will also discover that biology as a whole has connections with the brain and psychological behaviors of a human being as well.

In this section, you will be able to read about the connection of psychology and physical biology in terms of determining human personality and overall evolution.

One human personality can be seen as an amalgamation of the various traits and characteristics both physical and emotional. This makes up the overall evolutionary component of the human existence. Without that gazebo biology combined with the psychological traits of a person, a human being will not be able to exist and function well. Here are some psychological tenets that can directly affect human biology and personality.

Sigmund Freud's Psychosexual Theory of Personality and Evolution

According to Sigmund Freud, there are for psychosexual stages of growth in a normal human. These are as follows:

The Oral Stage

This is where a human infant has a tendency to do everything with his or her hands and mouth. The baby will do everything to have something to chew on no matter what it could be. This stage occurs from 0 to 3 months.

The Anal Stage

In this stage, the baby tends to feel the urge to touch his nether regions. Parents have to be careful regarding how they deal with their children during this stage because this is the stage where children who would like to explore their various orifices much more as an infant.

Phallic Stage

This is where the child learns to explore his sexuality even more. He learns how the reproductive organs work and how it relates to his personality. It is in the stage where he learns to pleasure himself sexually as a child.

Latency Stage

This is the phase where all sexual thoughts are tend to be repressed during childhood. It occurs from the ages 5 to 7 years old. It also lasts until the child reaches his puberty stage. This is where all the psychosexual thoughts developed in the early stages are apparently resolved and put under control by either the parents or the child himself.

Personality and Biology

In connection to biology, early medical studies yielded that personality is defined by the morphology of the brain. According to some scientists, the human intelligence and personality can directly be linked to the shape and size of the brain. This can further proven by the evolution of man according to Darwin.

However, modern studies have shown that Albert Einstein's brain had more gray matter than the average human, but is normal sized. Because of this, it was deemed that it is the amount of gray matter inside the brain that determines intelligence and the person's susceptibility to suggestion, not the size or shape.

Through years of development and evolution, the Homo Sapien has been able to gain sufficient intelligence that set them apart from their ape ancestors. Due to evolution as well, the human species were able to stand upright and learn to improve psychologically on their own.

Additionally, in the psychosexual stages of development, according to Freud the three main parts of the personality develop. The Id, Ego and the Super Ego. Freud linked personality to the biological and sexual desires of man.

In later years however, the emergence of other schools of thought regarding personality refuted this claim. Most of these experts say that it is not only sexual desire that fuels personality; it is the hormonal fluctuations and external environment that shape personality.

Conflicting Studies and Schools of Thought

According to Carl Jung, the expert who fostered the person-centered school of psychology and personality, the concept of personality is determined by various constructs from external sources such as family, society and other external factors that can affect the formation of personality.

Constructs can be defined as the individual concepts of people in regards to others. This is also known as the individual perception of one person in relation to another. According to Jung, these concepts of the brain determine the formation of personality as a whole.

Therefore, the formation of personality according to the same expert has nothing to do with the corporeal body. It has more to do with the metaphysical and mental aspects of existence. These conflicting schools of thought definitely are worth exploring with regards the quest to understand how significant biology is to the formation of personality. Learning all that you can about personality whether your own or another person's would be able to give you the tools to handle biological changes that may affect your personality more effectively.

The subsequent paragraphs will explain more about evolution and its role in the development of human biology as a whole. It is definitely something interesting to discuss and will be worth reading about as soon as possible.

Chapter 4 – Homeostasis

Understanding Homeostasis

This is the process of self-regulation that is carried on by the biological systems op maintain the required stability required for adjusting to the conditions that are considered optimal for its survival. If this state of homeostasis continues successfully then life will continue but then when this doesn't it can result in death or disaster. The stability that is attained due to homeostasis is usually that of dynamic equilibrium in which the change occurs continuously but it happens in conditions that are relatively uniform.

Additionally, homeostasis refers to the body's natural ability to achieve relaxation through internal regulation of temperature as well as other fluids and hormones inside the body. It is the physical vessel's way of maintaining regularity in functioning without medication intake as much as possible.

According to various sources, Homeostasis can also be defined as the internal regulation of various physical factors to maintain a certain sense of balance within the body. By achieving home use cases, the physical body will be able to produce the right amount of hormones that can induce happiness as well as other positive emotions that can make a person healthy all throughout his life.

Any system that has managed to reach a state of dynamic equilibrium would tend to reach such state in a steady manner while being capable of resisting the changes that can be caused by outside forces. When a system like this is disturbed then the built in regulatory system in the organism would respond to such deviations by making new changes and this process is referred to as feedback control. All the different processes of integration as well as coordination of the various functions whether facilitated by the electrical circuits or by the nervous and hormonal systems; these are all examples of homeostatic regulation.

For getting a better understanding of the manner in which the homeostatic regulations work let's take an example of a mechanical system like a thermostat or a temperature regulator for controlling the temperature of the room. The bimetallic strip of the thermostat is the vital part of the thermostat and this strip responds to any temperature change by either completing or disrupting the electric circuit. So, when the temperature of the room cools down, the circuit is completed, the furnace will work and the temperature would rise and vice versa. The biological systems of organisms are much more complex than any mechanical device and the regulators in the system are only roughly similar to those of a mechanical device. The aim of both the biological as well as the mechanical systems is the optimization of functions for sustaining themselves within the pre-described limits.

One instance of homeostasis in the body is that of temperature regulation, the normal body temperature usually ranges between 37°C, but there are different factors like hormones and even metabolic rate than can affect this value causing either high or low temperatures. The region of the human body responsible for controlling the temperature is the hypothalamus. The feedback regarding the body temperature is carried out through the body via the bloodstream to the brain and these results in compensatory adjustments by the system such as changes in the rate of breathing, level of blood sugar and metabolic rate. Heat can be lost from the human body due to activities like precipitation and heat exchange mechanisms. Heat loss can be reduced by insulation, decreased circulation of blood to the skin and various other things like the usage of warm clothes, shelter and external sources of heat such as heaters. The homeostatic plateau is the region between the high and the low body temperatures that is regarded as the normal temperature. If your body reaches either of the extremes the body starts taking corrective action on its own.

Examples Of Homeostasis

Homeostasis is the level of metabolic balance that is maintained by different body processes. There are different examples of homeostasis that take place in the human body, and learning more about these processes will help in getting a better understanding of how the body maintains its normal functions.

Our body is responsible for controlling the levels of acids and bases in the bloodstream. When the amount of the acidic components increases in your blood, then the body acidity also increases. When either the consumption or production of acidic compounds tends to increase or when the body can no longer eliminate any of the acidic compounds, then the acidity in the body will increase.

When the amount of alkaline compounds in the blood increases then the alkalinity of the body also increases. Acid-base balance in the blood is the balance between the alkalinity and acidity that's measured on the pH scale. This balance is controlled by the kidneys and lungs along with other buffer systems. Excess acids as well as bases are excreted by the kidneys. Kidney damage would result in the reduction of the ability of the kidneys to excrete those substances that cause pH imbalance in the body. The lungs control the pH level by the expulsion of carbon dioxide from the body. During exhalation the carbon dioxide is pushed out of the body and the pH of the blood changes depending upon the speed and depth of the breaths taken. This means that you can adjust your pH level in less than a minute by just getting your breathing under control. The buffer system helps in preventing any sudden pH imbalances because it consists of certain weak acids and bases that naturally occur in the body.

Controlling of the body temperature is also an example of Homeostasis in the body. The normal body temperature is around 98.6°F and when the body temperature falls below this mark, there will be serious affect on the body. It can cause muscle failure, loss of consciousness, break down of the central nervous system and in few cases it even results in death. The body temperature is controlled by either generating or expelling heat from the body.

The amount of blood sugar that is present in the blood stream is referred to as glucose concentration. The body makes use of glucose as a source of energy but having too much or too little of glucose will have serious complications because the body makes use of difference hormones for regulating this concentration of glucose in the body. Insulin helps in reducing the glucose levels whereas cortisol helps in elevating the glucose levels.

Our bones and teeth are made up of calcium, and they contain about 99% of calcium in the body. The other 1% is contained in the blood. Having too much or too little of calcium in the blood isn't good. When the blood calcium level decreases, the parathyroid glands help in restoring some balance. This gland is responsible for maintaining the level of calcium in the blood.

The body is required to regulate the level of fluid in the body for maintaining an internal environment. Hormones will help in regulating the balance by helping in excretion or retention of fluid. When the fluid volume in the body is less than the kidneys start retaining fluid and the urine output also decreases. If the body has too much fluid then the opposite of this takes place.

Different Factors That May Affect Internal Balance or Homeostasis

Many possible factors can affect our internal balance both in the positive and negative sense. This section will give you an ad as to what those factors are and how you can tip things to your physical favor when you do experience problems with your internal balance in the near future.

Lack of the Necessary Hormones

When the body is unable to produce certain hormones that it needs to function well, a human can experience many possible side effects. For example, if the pituitary gland of the brain fails to produce the necessary growth hormones, internal balance can be disrupted and therefore cause a person's height to become stunted. As a result, an affected individual can become dwarfed or lacking in height. Inversely, excessive release of growth hormones from the pituitary gland can lead to gigantism or severe enlargement of the body.

In women, a significant lack of hormones especially during menopause can certainly produce different reactions from the body. A lack of estrogen and progesterone in women can result to hot flashes especially in the early stages of menopause.

In addition to this, the lack of hormones can also affect the elasticity of the skin, which can result to dryness and sagging. Losing hormones can also affect their ability to bear children after a certain age. As a matter of fact, this is the most prominent effect of menopause to some women. It is also the reason why most of them opt to undergo hormonal replacement therapy particularly be on menopause so that they will not end up experiencing any of the physical side effects of losing the hormones.

Stress and Tension

We all know that a healthy amount of stress and tension can be good for the body, especially when it comes to developing physical strength and endurance as well as muscular tonality. However, excessive stress can also have negative effects on the physical form. Too much stress can lead to excessive sweating, increased temperature and possible aches and pains all over the body. If this happens, the human vessel will certainly not be able to function well in his everyday life.

Aside from the physical effects of the stress and tension, you can also be affected by stressed emotionally. If this happens, your body and will not be able to regulate some of the natural hormones that it needs to function well internally as well as externally. For example, if you are affected by stress emotionally, chances are you will not be able to eat well.

This failure to eat can lead to various physical complications such as malnutrition and a fluctuating blood pressure. If this happens, the internal balance of your body or homeostasis will be put in jeopardy. Whether you eat too much or you eat too little, you will be able to experience severe side effects by doing either of the two.

Further research has shown that everything which is done in excess both physically and emotionally can directly affect the internal balance of the body. For example, overheating will certainly lead to complications in the blood pressure as well as the cardiac health of an individual. It can lead to negative medical conditions such as the following:

- High blood pressure

- Diabetes

- Kidney failure

- Heart problems

- Obesity

Experiencing these different types of physical ailments can definitely throw off the internal balance or homeostasis of the body. This is why it is important for you to maintain a balanced diet. Eating the right kinds of food will definitely help improve your physical health as well as your disposition in life.

You can also deal with stress and emotional problems through meditation. Doing some breathing exercises will help you relax and take your mind off of problems that may affect your internal organs and physical body in the process.

Do you ever wonder why some people ask you to take a deep breath whenever you feel highly emotional? This is because breathing deeply will allow you to inhale oxygen that helps regulate significant blood flow into the body. By allowing oxygen you get into your body effectively, you will be able to foster a good environment for your heart and other parts of your corporeal form. This is why it is important for you to do breathing exercises.

The Significance of Physical Activities to Homeostasis

You can also achieve the internal balance that you have been looking for by doing some physical exercises and activities around the house. Being physical around the house can help your muscles get used to the tension that physical movement brings. If you move a lot, your body will be able to handle the physical impact that certain events in the environment can bring. It will learn to develop a certain tolerance to foreign bodies that can be a cause for various illnesses.

In addition to this, physical activities can help regulate the internal hormones that the body requires a function daily. By having a regular exercise routine, you will be able to maintain the peak of your appearance as well as regulate your own stress levels without having to resort to medication.

Doing some form of exercise is only one way to achieve homeostasis. There are other non-physical ways to deal with this issue in the future. The next few paragraphs will deal with the non-physical side of achieving homeostasis.

The Advantage of Natural Healing To Homeostasis and Biology

As mentioned earlier, you do not need to resort to purchasing mind-altering drugs just to achieve homeostasis. There are many natural ways to do so. This section will deal with some of the options that are available when it comes to natural healing that can also be advantageous in terms of maintaining homeostasis inside the body. Here they are as follows:

Initially, you just have to learn to eat right. If you want to avoid various physical ailments that may end up causing you to go to the hospital on a regular basis, changing your diet is one way to avert the problem. You should do your best to add vegetables, fruits, dairy and other healthy types of food to your diet. If you're not sure what to eat, consult a nutritionist are a dietitian to learn more about the food chart and how it can work to your advantage in the future.

Additionally, you have to avoid stressful situations as much as possible. If you think that you easily are stressed, you should try to determine what those stressors are and how you can avoid them. This way, you will not end up experiencing the debilitating side effects of stress in the long run. It will also help you maintain internal balance as well.

You should always keep the importance of sleep in mind. Having a good night's sleep will be able to help your body recover from the daily grind of physical activity. It will also help the brain relax and gather more oxygen for the next day. It is for this reason that you should always try to get 8 hours of sleep during the evenings. According to researchers, it is during certain times at night that the body undergoes internal reparations. So if you do not sleep at night, you will therefore end up disrupting the internal balance because you are not giving the body time to recover from the stress of the day.

Drinking lots of water will also allow your body to achieve internal balance. Since the body is made up of 70% water, sweating it off after a bout of exercise can definitely make you lose a lot of energy. Energy is important to achieve great balance.

You can definitely replenish the energy by drinking lots of water. It is recommended that you drink 8 to 10 glasses of water a day to maintain optimum corporeal health.

The Different Complications of Dehydration

Aside from helping you achieve internal balance, having a substantial supply of water inside the body will allow you to prevent detrimental illnesses from occurring. One of these illnesses would be dehydration. Dehydration can lead to other ailments and complications as well. Some of these complications are as follows:

- Heat stroke

- Swelling of the brain

- Seizures

- Low blood volume

Having insufficient water in the body can also lead to a coma and potential death. This is why it is important for you to drink lots of water. Not having enough water can lead to biological complications that will be detrimental to your functioning as a normal human being.

Brain Swelling

I would like you to focus your attention to the swelling of the brain. Not having sufficient water supply to replenish your body can actually affect the multitude of brain cells that you have. Brain swelling can be caused by the physical body trying to supply the brain cells with the remaining amounts of fluid that it has left. If this happens, it can create pressure inside the brain and cause the cells to rupture. This will eventually lead to brain damage, which can lead to different sets of related problems in the end.

These are just some of the reasons why you need to make it a point to drink as much water as you can. If you do not want to have additional problems down the line, drinking lots of water will definitely help you along the way.

It is also important to remember that you should drink pure water as much as possible. This way, you will be able to avoid impurities entering your bloodstream and potentially affect your body negatively.

Cold vs. Hot Water

Depending on the condition of your body, drinking cold or hot water can be both good and bad for you. Drinking cold water can be good at keeping yourself refreshed during a hot summer day. However, drinking too much cold water can result in various problems with your throat. This is why it is important for you to achieve balance. You should learn to listen to your body. If you feel tired, take a rest. Do not force yourself to do anything physical if you are not feeling well. Exerting too much effort can also upset the balance of your internal constitution. You should learn to find a middle ground for your body. This way, you will not end up collapsing are losing more energy than you really should in the near future.

Homeostasis and Losing Weight

Furthermore, achieving internal homeostasis can help you lose weight. Because you are well balanced inside and out, you will not feel the need to eat up at odd times during the day. Because of this coupled with physical activity, you will end up having the right amount of calories inside your body which will render you healthy and illness free.

Always remember to have the right reasons for losing weight. Do it because you want to be healthy and not because you want to please other people. If you end up dieting for the wrong reasons, this will add to your stress levels significantly and eventually cause severe imbalances in the body. This will then break the homeostatic process that you have been going through down the line.

Keep in mind that it is all about balance. If you are not able to achieve physical balance, try to achieve emotional balance first. If you are at peace emotionally, I am sure that everything else will surely follow. You just need to believe that you can do it and you'll end up succeeding in everything that you do.

In addition, when trying to lose weight, it is important not to starve yourself. By doing this, you will deprive your body of the nutrients that it needs to function well. You will also end up tampering with your own metabolism, which is a significant part of homeostatic functioning inside the body as mentioned earlier.

If you want to achieve homeostasis, you should be able to align the physical, emotional and spiritual needs of your body into one complimentary system. This way, you will not end up doing too much or too little to improve yourself. You will also be able to avoid additional illnesses if you are able to find balance in many ways.

Losing weight can be very difficult for some, but once you find the right formula and achieve enough discipline in yourself, it will definitely work to your advantage for sure. Again, it is all about finding the right balance between what you should and should not do. Once you achieve this, everything else will surely follow.

Robert Meeks

Chapter 5 – Energy

The human body, like the body of any other living creature is an energy conversion machine. Energy conversion means that the chemical energy that is found in food is converted into work or thermal energy. It can also mean the conversion of chemical energy in the fatty tissue. The largest fraction would go to thermal energy, but this proportion varies depending upon the kind of physical activity that you involve yourself in. The fraction of energy that goes into each form also depends on the food that you consume along with the kind of physical activity that you have engaged yourself in. If the consumption of food is more than what is required for work and for staying warm then the rest would be stored as body fat.

All the bodily functions performed require energy. The bodily functions can range from thinking to lifting weights or going for a run. All the small muscle activities that go along with quiet activity such as sleeping to scratching your head are converted into thermal energy and so are the activities that aren't visible like the muscle actions of lungs, heart and even then digestive tract. Shivering is actually the automatic response of the body that is generated so that some thermal energy would be produced for increasing the low body temperature. When you shiver, the muscles are rubbed against one another and this generates heat. Vital organs like kidney and liver consume a considerable share of the energy for carrying on their functions, but a total of 25% all the energy in your body is made use of for maintaining the electrical potentials present in all the living cells.

Basal metabolic rate is the rate at which the body makes use of food for sustaining itself and also for all the various activities that it undertakes. The total rate of energy conversion that is undertaken when the person is resting is referred to as the BMR and this is divided among all the other systems present in the body. The largest fraction of all the energy that is produced in the body is diverted towards the function that is carried out by the liver, kidneys, spleen and the brain. The consumption of energy by the skeletal muscles as well as the heart increases while exercising. Almost 75% of the energy that you burn in a day goes towards the performance of these functions alone. Useful work is the work that is done by a person in the outside world. For instance, lifting weights can be considered as useful work. Useful work requires that the body exerts a force through a distance on the outside world and this excludes internal work such as the pumping of blood that's done by the heart and so on. Useful work includes anything that exerts force on the outside world. So, running, jogging or even simply walking can be included in this category of work.

Energy exists in different forms in nature and this can neither be created nor is it capable of being destroyed. Energy is just capable of being transformed into one form from the other. Examples of energy would be kinetic energy, potential energy, thermal energy, electromagnetic energy, chemical energy, nuclear energy and so on. Energy is usually expressed in terms of either joules or ergs. In biology, energy is usually stored by the cells (building blocks of matter) in the form of biomolecules like carbohydrates and lipids. The energy that is released by these molecules when they have been oxidized during the process of respiration is then carried and transported to the various energy carrying molecules that are referred to as ATP.

The human body is capable of surviving on whatever food that is available. The ability of human beings to survive on a variety of diets is one of the reasons why this species is adaptable to any given condition and can survive on any source of food. Calorie intake these days has increased and unhealthy food habits seem to be the norm of the day. Therefore, it is really important to have an idea of what macronutrients are and the reasons why it is important to make better food choices for leading a healthy life. Each morsel of food that you consume is broken down into several smaller parts so that the body can make use of them. It is essential for survival that a regular diet should consist of proteins, carbohydrates and fats. If all these three are present in a normal diet, then the body will make use of all of fats and carbohydrates for generating energy and proteins would be made use of for making muscles, producing hormones and other basic biological equipment.

Proteins that are present in food are broken down into smaller parts, which are referred to as amino acids and these acids are used to build building new proteins that will be employed for specific functions like acting as a catalyst for chemical reactions, transportation of biological molecules and also for the generation of energy when there is a shortage of fast and carbohydrates. More than half the energy requirements of the body are met by fats. The fat that's present in the food is broken down into fatty acids that are capable of travelling throughout the body and they are taken in by those cells that need energy. Fats are stored in the form of triglycerides in the fat cells when they aren't made use of right away. The carbohydrates that you consume can be stored in only a certain manner in your body and these are the first nutrients that your body makes use of for the generation of energy. Carbs are broken down into either glucose or sugar by the human body and these are made use of for providing energy for carrying out various functions.

Cell and Energy: The Correlation

Cells also have all-important energy that is necessary for them to survive. Without cellular energy, these organisms will not be able to function normally. As mentioned earlier, the energy comes from a combination of natural chemicals produced by the body through the various nutrients that are taken in every day. It is through this idea as well that the body is able to produce the necessary energy to allow the organs to work properly without fail. Without the proper nutrients, this else will not be able to produce energy and in turn, render the body or some organs unable to function.

To provide the body and cells with the chemicals that it needs to produce energy, eating the right kind of food is necessary. Carbohydrates, vitamins and minerals are just some of the most important nutrients that the body needs to facilitate cellular regeneration normally. So if I were you, I will definitely go ahead and eat as much pasta and rice you would like if you require a lot of energy for the day. Here are some other examples of "go foods" that you need to include in your diet as soon as possible.

- Cereals

- Bread

- Meat

As mentioned earlier, you should include this in your diet if you don't want to lose energy during the day.

Potential and Kinetic Energy

There are two different kinds of energy. These are known as the potential and kinetic energy. The former refers to start energy that is not in use within the body. It is transformed into kinetic energy once it is released to perform various tasks.

Once energy is used up and released, you would need to give actually replace it to keep the body functioning well. This is where proper food intake is taken into consideration. You would need to ingest full and that is rich in carbohydrates and other nutritious elements to replenish those that you have lost. Here are some examples of carbohydrate rich food groups that you should keep in mind.

- Rice

- Meat

- Fruits and vegetables

- Pasta

If you are able to add any of these to your diet, then you are well on your way to having more energy than you can possibly imagine. As mentioned in the earlier text, it is important to maintain a balanced diet so that you do not end up losing too many nutrients that are of equal importance when it comes to building up your body's physical strength and stamina.

You also have to take in a form that has sufficient amounts of protein and calcium as well as fiber so that you end up having strong bones and muscles to complement the power that your body will gain once it has all the energy that it needs.

5 Effective Ways to Conserve Energy through Meditation

If you live an active lifestyle, it is inevitable that you are going to have to spend a considerable amount of energy depending on your daily tasks. However, what if I told you that there was a way for you to conserve your energy without having to spend more than you already do? This section of the book will discuss some of the most effective ways to conserve energy with meditation.

Silence

This is one of the most effective ways for you to conserve your bodily energy. In order to make sure that you are able to achieve inner silence, you should be able to practice at work silence as well. This will help you conserve sufficient energy for other activities during the day. It will also help you organize your thoughts to foster a better functioning brain in the process.

Abstinence

If you really want to conserve energy, it is important that you are able to abstain from any kind of physical activity for the near future. Stay away from strenuous activities that might make you more energy in the future. In this same vein, it would also help you to abstain from sexual activity to conserve more potential energy in the future.

If in case, you do need to move around and do something. Make sure that it is something productive. Exerting energy on something that you like will not feel like work at all. It will take your mind off your problems and you will eventually feel rejuvenated in the end.

Try to Avoid Negative Thoughts

If you really want to conserve energy, try not to think too much. Physical exertion is not the only culprit when it comes to severe energy loss. Sometimes, even if you are not doing anything, you can still lose a significant amount of energy if you keep on worrying about certain situations that are out of your control.

The best way to deal with negative thoughts is to make sure that you are able to find out the root cause of your negativity. For example, if you are stressed about work, try to devise a way for you to be able to relax while working still. Face your problems head on and you will be able to find solutions sooner than you think. As much as possible, try to be methodical in your problem solving, this way, you will be able to work out your problems without feeling stressed at all.

If in case you cannot find a solution to your problem right away, it would help you to seek advice from other people. Sometimes you just have to see our problems through a new set of eyes for you to resolve them effectively. This way, you won't have to worry too much about what is happening around you.

Live with Passion

Going back to number 2, try to live as passionately as you can. If you're able to live with passion, you won't feel like you're being drained of your energy. Sometimes, it is all in the mind. When you think that you are being drained of your energy, chances are your body will be able to feel it a thousand times over. Remember that your mind is a powerful thing. If you think about something long enough, it will eventually come true. This is why you have to be positive and fill your body with good thoughts so that you would be able to function well enough to go through your daily activities in the long run.

Lastly, you have to do some breathing exercises in order to replenish your energy. Doing yoga and quiet meditation will help you become more attuned to your body. You will be able to incur more oxygen as well if you are able to breathe normally and in a more relaxed manner as well.

These are just some of the many ways for you to conserve your energy. Always remember that your body is the temple of your soul and if you don't have enough energy to use it, then you won't be able to fulfill all that you want to do in your life. This is where biology and meta-physicality meet. Aside from ingesting good food as mentioned earlier, you have to have a great disposition as well. This way, your mind and body will align and you will become an ultimate human entity for sure.

Energy and Your Environment: How Do Biological Interactions Affect Energy?

Energy and the environment play an important role in fostering a good and healthy biological existence. However, sometimes due to certain events, we tend to fail ourselves and fall ill. However, you can still have enough energy to do the most important things in life even if you are sick. You do not have to completely stop working in order to successfully rehabilitate yourself.

Here are some tips for you to learn how to work with what you have in terms of conserving energy while maintaining a safe environment for your body as well.

Slow Down

First, what you need to do is to learn to pace yourself. Do not try to do everything at once. Work with your body in terms of giving it something that it can do without over exerting yourself. This way, you will be able to conserve your energy while still being able to do the work that needs to be done during the day.

Ask for Help

Secondly, you have to make sure that you are able to delegate the things that you won't be able to do on your own. Your body will thank you for allowing it to relax and rest for a while by having somebody else do things for you.

You will be able to conserve your energy for other activities if you are able to delegate your tasks accordingly. Biologically speaking, your body works in a similar manner in terms of delegation. If one of the cells is not functioning well, the brain will do everything to send as much blood as it can to other better functioning organ so that it can compensate for the one part of your body that does not work well. This will allow that part of your body to rest and gain more energy through recharging.

Listen to Your Body

If you feel unwell, take a break. Do not force yourself to do anything physical if you feel that you cannot deliver. Forcing your body to exert more effort when it can't clearly function well anymore because of disease will not bode well for your recovery.

One clue about your body is not ready for work would be pain. If you feel a certain amount of pain when you move, chances are that your body is still recovering from trauma. So you have to be very careful about any physical activities that you may have in mind in the near future. You should let your body heal first before you can do anything drastic physically.

That being said, you should also maintain a certain sense of mobility during the recovery phase. You ought to do this so that your muscles and internal organs will not end up atrophying and malfunctioning in the end. This way, you will have a faster recovery time if you maintain a certain amount of mobility in place despite your sickness.

Overall, our environment dictates how we can spend our energy. This is why you have to surround yourself with physical activities as much as possible so that you would not end up being a couch potato all your life. However, with the advancements in technology, physical activity has ever so slowly been losing its allure. This is why you have to make it a point to move around the house even if you do not want to. Otherwise, you'll feel drained all the time.

Have you ever experienced feeling so tired even if you are just lying down in your bed? Or feeling pain even if you're not doing anything physically exhausting? This is because the machines are draining you of your energy through fostering lesser brain activity. If your spend your days just staring at the computer all day long without actually doing anything substantial, chances are you'll feel tired after just a few minutes.

Additionally, your body will be using a certain posture that might be detrimental to your physical health in the long run. For example, if you sit too long, your hips will end up being frozen into a certain position. If this happens, you will end up with a stiff pelvic bone, which will affect your gait significantly. To prevent this, you should be able to take periodic breaks and stand up as many times you can between your stationery work.

This way, your muscles will get used to having tension in a regular manner and therefore not turn numb in the long run. You have to remember that motion is very important when it comes to blood circulation. So if you want your legs to be in great shape, it would help to keep them moving as much as possible. You will not regret doing so for sure.

Closing Statements

These are additional ways for you to conserve energy if in case you do fall ill. Sometimes, we just have to accept the fact that we need to rest for a time in order for the body to recover effectively. However, if you are still willing to work despite your illness, you just have to follow the steps above to make sure that you are able to deliver and become even more productive in the future.

I hope that you will continue to learn more about your body in the future and make sure that you are able to adjust your physical activities to your energy level. Ultimately, resting your body is one activity that can help you fully recover your energy.

Robert Meeks

Chapter 6 – Genetics

The Study Of Heredity

Genetics is the study of the heritable traits and their transmission from parents to the next generation that is their offspring. Traits tend to be similar in families and this is a general observation. Only during the mid-nineteenth century did people realize the larger implications that genetic inheritance would have on the species as a whole.

In 1858, both Charles Darwin and Alfred Russell Wallace had announced their theory regarding natural selection. According to Darwin's theory, all individuals' populations tend to produce more offspring than are necessary to replace themselves. If every individual alive would want to live and then produce further offspring then the entire population would collapse. Due to the restriction of the availability of resources overpopulation would result in competition for these limited resources. Darwin had made an observation that it is very rare for two individuals to be exactly same. He explained that these natural variations that exist among individuals is the reason for natural selection and those individuals who have been born with variations that will be advantageous for them and also confer the chances of obtaining resources or mates tend to have greater chances of passing these on to their offspring. Individuals who are born with different variations are less likely to reproduce.

Darwin knew that it was natural selection that leads to the creation of new variations of traits in the population or even the creation of entirely new species, but he wasn't able to explain the origin of these variations. He was completely unaware of the work that was being carried on by a monk named Gregor Mendel in the field of genetics. In the year 1866, Gregor Mendel had published all the results that he had managed to gather after years on experimentation of breeding pea plants. He observed that both parents need to pass on discrete physical factors that would help in transmitting their traits to their offspring at the time of conception.

An individual tends to receive from each of his parents one such unit that would define the traits of that individual. The principle that Mendel had come up with explains that the traits that a child inherits aren't necessarily a blend of the traits of their father or mother. When an offspring tends to receive the opposing forms of the same trait then the more dominant trait would be apparent in that individual whereas the recessive trait, though not apparent, would still be a part of the genetic makeup of such an individual and can be passed on to their offspring in the future.

Mendel's experiments had shown that when the sex cells are formed, there are various factors that would accompany each of these traits that the individual would inherit from its parents and these are then separated into different sex cells. When the sex cells come together during conception, the resulting offspring would contain at least two factors that are transferred from each trait. One of these traits would be inherited from the mother and the other trait from the father. Mendel made use of the laws of probability for demonstrating the fact that when the sex cells are formed it is just a matter of mere chance regarding the factor that would be integrated into a particular sperm or an egg.

The concept of simple dominance is not sufficient for explaining all the traits, and there some cases of co-dominance. This is when both the forms of the traits are equally expressed in the offspring. Incomplete dominance is a result of the blending of traits. In case there are multiple alleles, then there are only two probable ways in which the given gene can be expressed in the offspring. Most of the expressed traits, such as the different variations in skin tones in human beings, are influenced by different genes that are all acting on the same trait.

Every gene that tends to act on a particular trait can have many alleles and even the environmental interaction would have an effect on the genetic makeup of an individual. This is the reason why sexual reproduction happens to be the biggest contributor towards all the genetic variation that exists among individuals of the same species. It was only during the twentieth century when the scientists realized that it would help if they combine the ideas of natural selection with genetics that they would actually make considerable progress in their study of understanding the different organisms present on earth.

Scientists had realized that the molecular constitution of genes must have a provision for including genetic information that is capable of being copied efficiently. Every cell that is present in the human body comes with an inbuilt instruction manual that would tell them how and when they need to build proteins (the building blocks of body structure) and the manner in which they would be responsible for different chemical reactions that take place in the body. It was in the year 1958, which James Watson and Francis Crick had explained about the structure of a DNA molecule and they explained how this made use of the information stored in the cells for the production of proteins. This information is stored in the nucleus of the cells.

Every time a cell needs to divide all this information stored in it needs to be transferred to the daughter cell and every time all this information of DNA is copied, then there are some inevitable changes that take place. Most of these changes are immediately discovered repaired but there are times when the alteration wouldn't have been repaired and it results in an altered protein. The chances of an altered protein to function normally aren't high and the genetic disorders are the result of malfunctioning proteins that have a negative effect on the organism. The instances where the altered protein is functioning better than the original trait are extremely rare and when this happens they provide a survival advantage. This happens to be a source of genetic variations that we see. There is another source that results in genetic variation and this is genetic flow. This means that new alleles have been added to the population. Migration can also cause this, where new individuals who belong to the same species have entered the population. Environmental conditions that would have prevailed in their previous home would have been different from the territory they have migrated to and therefore their alleles would be different from the ones of those present in the new territory. And when they interbreed with the host population, that's when new forms of traits would be introduced to the population and the alleles that are favorable would spread through the population.

Genetic drift is referred to that change in the allele frequency that is caused randomly and not due to the pressures of selection. Mendel had described that alleles are stored in a random manner in the sex cells. This means that at times it could so happen that both the parents might have contributed the same allele for a given gene and when such an offspring produces; only one of the inherited traits can be transferred to its offspring. Genetic drift is responsible for causing remarkable changes in the entire population over the period of only a few generations if the population size is small. This reduces any genetic variations that might exist in the population. When this takes place, the chances of the decimation of the entire population due to a common cause are greater because of the lack of any genetic variations.

Genetics 101

Genetics can be defined as the study of the human genome, heredity as well as the variations in genetic code. It is generally considered as a biological field of study yet it intersects with many of the other Life Sciences in education and the further study of information systems.

As related to biology, the study of Information Systems deals with the isolation of any organized network utilized to gather information about the environment. In this case, genetics and its very essence would be the study of how the body is able to gather information and collected into a multitude of DNA strands that determined the innermost traits of a single human being.

The brain, in particular, is considered as the most innovative supercomputer in the world. Therefore, it is safe to say that studying genetics will help us understand the internal workings of the mind and other parts of the body in the long run.

Every individual has got about 20,000 genes altogether and we all share the same set of genes. The differences that we notice in individuals are due to slight variations that take place in the genes. For instance, a person who has red hair does not have a gene that gives red hair and the one who has brown hair doesn't have the brown hair gene. Instead everyone has got genes for hair color and depending upon the version of the gene, the hair color of the person would be determined. There are about 50 trillion tiny cells that are present in your body and every single one of these cells contains the complete set of instructions that define who you are. These cells are the building blocks of every individual.

All these instructions are coded in the DNA and the DNA is a ladder like molecule. Every rung of this ladder has two units that are interlocked and these are referred to as bases. Four letters of the alphabet are made use of for referring to this. And the letters are A, T, G and C. A is always paired with T and G with C. chromosomes are the pieces of the long molecules that consist the DNA. Human beings have 23 pairs of chromosomes and this number varies from one organism to another. For instance, chimps have 24 pairs of chromosomes. The number of chromosome does not dictate the complexity of the organism.

These chromosomes are then further divided into shorter segments that form a part of the DNA and are referred to as DNA. Think of your DNA as a cookbook, and all your genes would be the recipes in the book. The genes would dictate the functioning of the cells in your body and the traits that your cells are supposed to exhibit. Cells in your body make use of all the recipes that are written in your genes for the production of proteins in a manner similar to following a recipe given in a cookbook. Proteins are responsible for most of the work that takes place in your cells as well as your body. There are some cells that would help in deterring the shape and structure of cells whereas there are others that would help with various biological processes like digestion or supply of oxygen in the body. There are different types of cells; there are brain cells, blood cells, skin cells, liver cells and so on. But then how do these cells know that they are supposed to make an eye and not a toe? Well, the answer is in the complex system of genetic switches. There are master genes that are responsible for turning on the other genes and also for making sure that the right proteins are being produced by the right cells. An existing cell will have to divide into two cells for it to create a new cell. But before it can do so, it will need to make a copy of its DNA for the new cells so that both the cells will have a complete set of instructions for themselves. While the process of copying of DNA takes place, there might be some mistakes that can occur and this changes the DNA in certain locations referred to as SNPs or single nucleotide polymorphisms. When SNPs are generated this creates biological variation between individuals and the recipes for protein construction will also change. These differences that have been caused can in turn affect a variety of things like the appearance of the individual, heir traits or even susceptibility towards any particular disease and so on.

DNA is passed on from parent to a child, so you have inherited your SNPs from your parent, and your child will obtain them from you. You will be a match to your siblings, aunts, cousins, uncles and grandparents at many of these particular SNPs that you have inherited. However, you would find only a few matches to your distant relatives. The SNPs can be used to determine your proximity to your relatives as well. Every human being is made of 23 pairs of chromosomes; one of each pair comes from each of your parents. The two chromosomes of a pair contain the same gene, with minor differences; only the sex chromosome contains different genes. The sex chromosome is unique, and you might not always end up with a matching pair. Females would have two X chromosomes, whereas males would have X and Y-chromosomes. Mothers always pass on the X chromosome, whereas the father might pass on an X or Y chromosome. The chromosome that's passed on from the father would determine the sex of the child.

Genetics and Biology

Biologically speaking, the genes or Deoxyribonucleic Acid can be found inside every living cell within the body. This is how sexuality, physical appearance, and many other external traits can be determined. This is why it is important to learn more about genetics because, without genetics, we will not be able to determine their innermost connections of human life.

Genetics can determine the physical appearance as well as arguably some personality traits of a human being. Physically, it can determine the color of your eyes or your height and skin tone. It can also determine paternity, which is why most people who are trying to confirm someone's identity research to DNA testing to be able to find out if they have any similarities in the genome pattern.

Genetic representation works in such a way that chromosomes combine in various correlations to one another. A different combination of chromosomes will lead to a variation in the development of a child. This is why missing one chromosome in the combination and the creation can lead to dire consequences for the child. Here are some of the various conditions that are linked to chromosomal aberrations:

Down Syndrome

A delay in physical growth, as well as a stark difference in terms of the morphology of the face and other physical features in the child, characterizes this genetic disorder. This particular disorder can also affect the brain patterns as well as the cognitive functions of an afflicted individual. People with Down Syndrome can be trained to live as normally as possible, there's still no known cure for this condition. The cause of the condition also is generally unknown. The development of the extra chromosome happens by chance apparently. This means that the parents of a child with Down Syndrome can be normal and still have a child afflicted with the condition.

The Chimera

The human chimera phenomenon can happen if an individual develops two sets of DNA. This process can happen in-utero if an infant happens to absorb the genetic material of his unborn twin. Because of this, any individuals that may have this condition can develop various genetic aberrations such as skin pigmentation, differently colored eyes and many others. This usually occurs in animals as well.

Albinism

This is a condition where the afflicted individual has an absence of skin pigment caused by the non-production of an internal body substance known as melanin. Being a genetic disorder, there is also no known cure for this condition.

These are just some of them most common genetic maladies that medical experts have encountered through the years. Some experts also consider hermaphrodites as a genetic condition. This occurs in various animal species as well as humans.

The Importance of Studying Genetics

This study of genetics is important because it will help people determine their own identities in the long run. It will help them find out who they really are in the grand scientific design. Genetics can also help in the advancement of medical science in terms of gathering data about certain human traits that the experts would want to explore. It will definitely give you a chance to figure out how your body works on a much more minute scale.

Various Genetic Phenomena

Many internal processes are still a mystery to the experts in the field of science. Genetics can help them discover further information about these conditions. They will be able to answer questions like how some animals can change sexual organs after a considerable amount of growth.

This particular scientific field can also shed some light on personality determinants as revealed by the genetic makeup. There are also some controversial topics that have been linked to heredity and genetics. One, in particular, has garnered a lot of attention in terms of dissertations and studies: The topic of homosexuality.

Experts are still debating whether this particular type of orientation occurs partly due to genetics or is it because of the environment. Various studies have proven both schools of thought with regards this issue. However, because of the ever-changing condition of the human race, new questions about the validity of the known findings still come up from time to time.

This is why genetics is a distinct field of study because it allows the human mind to peer into unchartered territory.

Conclusion

Biology is a significant branch of science that will allow you to explore life in a more in-depth manner. With all of the vast information about cells, evolution, and energy, it is definitely one of the most interesting subjects they teach in school. This is why I encourage you to continue exploring biology as a science. This book is just the beginning of your journey. There are many other sources out there that can help you understand body, and biology in general. All you need to do is to look and you will surely find the answers to your questions.

These are just some of the many significant pieces of information that you can encounter about biology in relation to the different fields of study. The human body is fascinating and complex. This is exactly the reason why most people are interested in learning all they can about the human body and how it works. Whether it is about reproduction or genetics, no one can deny the significance of this particular field when it comes to scientific studies.

We have covered all the basic concepts that you need to know about biology that's the science of life. In this book we have taken a look at some of the important aspects of biology like evolution, genetics, homeostasis and energy. Knowing more about these concepts will definitely give you a fresher perspective towards life and also help you understand why your body functions the way it does.

Biology might have seemed like an intimating subject, but once you break it down to its basic elements, biology is not that scary and you might have realized this by now. It is all about trying to understand the origin of life and its development.

So, I would like to thank you once again for choosing this book and hopefully it has proven to be an informative read.

Robert Meeks

And lastly, I'd like to ask you for a favor; would you be kind enough to leave a comment for this book on Amazon by posting a review? It'd be greatly appreciated!

Robert Meeks

www.ingramcontent.com/pod-product-compliance
Lightning Source LLC
Chambersburg PA
CBHW060410190526
45169CB00002B/833